Automation Engineers

Discussions Promotions Jobs Members Search

Do you agree with the prediction that PCs will replace PLCs in certain Advanced Automation applications?

Enrique (Hank) Vazquez
★★★★★ Your Plant Automation Concierge! ✔Control System Integrator ✔PLC Programming ✔PID Tuning ✔Consultant Engineer
Top Contributor

http://www.thomasnet.com/journals/machining/the-future-of-advanced-automation-in-precision-machining/

The Future of Advanced Automation in Precision Machining thomasnet.com
Increased software usability and hardware versatility has opened up automation applications to lower-volume production as well as part families.

Like (3) • Comment (26) • Share • Unfollow • 4 days ago

Paul Bennett IEng MIET
Systems Engineer at HIDECS Consultancy

Paul

The PC as we know it, no. A more capable class of processor certainly. There are already plenty of machines with bespoke controls that feature better than Pentium Class processors (often more than one). They also require high end Analogue and Digital Sensor Systems with very accurate machining in their construction.

Unlike • Reply privately • Flag as inappropriate • 4 days ago

👍 Enrique (Hank) Vazquez likes this

Joan Murt Llumà
R+D manager in Technologies For Advanced Manufacturing and Robotics
Top Contributor

Joan

Yes, of course.

There's no possible discussion here: PC's are much more powerful than PLC.

It has no sense using specific hardware that is much more expensive than what it should when you could use a powerful PC.

I've been using them (Beckhoff PC's and software) for 15 years and I've not had any problem.

There are things that can't be done with conventional PLC's but are a breeze with a PC.

15 years ago was possible to ask this as the PC control technology was still young and not very common. Today it has no sense asking that.

Cheaper, faster, it gives you freedom...

Unlike • Reply privately • Flag as inappropriate • 4 days ago

👍 Enrique (Hank) Vazquez likes this

Duane

Duane Klein
Software Product Manager at T&H

No. This has been tried before. Each new generation of PC comes out and someone says, "This time it will be different." A PLC is a dedicated control device. In manufacturing, you need something that has a single focus so as to insure quality, performance, and reliability. You could take a PC, make it compact, hardened, and with a dedicated OS so as to perform in this fashion, but then you would have recreated a PLC.

Unlike • Reply privately • Flag as inappropriate • 4 days ago

👍 Andrzej Czechowski, Enrique (Hank) Vazquez and 2 others like this

Joan

Joan Murt Llumà
R+D manager in Technologies For Advanced Manufacturing and Robotics
Top Contributor

I've made machines with up to 26 CNC axis with only one PC, Microsoft Visual C++ to create the HMI and Beckhoff TWinCAT to program the PC based PLC and CNC.

One PC can control more than 256 CNC axis, control up to 10 PLC tasks and show a rich HMI made in any high level language (Visual C++, .NET...).

The current power of the PC's out there is much bigger than any PLC manufactured. You can always choose between a non-powerful dedicated device or a super-powerful non dedicated device.

It must not be different, it is simply a different concept...

When you buy a PC you get a machine that is capable to calculate at a defined speed with a large amount of memory available.
When using it in industrial applications you can use that power to make what you need.

Unlike • Reply privately • Flag as inappropriate • 3 days ago

Enrique (Hank) Vazquez likes this

Paul

Paul Bennett IEng MIET
Systems Engineer at HIDECS Consultancy

@Joan; While you may think that PC's are more powerful than PLC's the PC (as a personal computer) is not suitable for machine control except as perhaps just the HMI in the control room. However, some of the PLC's I know of are using high end chips that are more powerful than some PC's have got and multiple such chips at that.
Developing machine controls requires attention to the Safety and Security aspects in order to meet the requirements of legislation and standards that apply to such equipment. As Duane observed, if you make the necessary ammendments to a PC to comply with such regulations and standards, for the purposes and controlling plant or machinery, then you are recreating a PLC anyway. Have a look at IEC61508, IEC61511, IEC61513, IEC62061, ISO13849 and ISO14118 for what is expected of a system. Not the sort of thing to be entrusted to a normal PC.

Unlike • Reply privately • Flag as inappropriate • 3 days ago

Enrique (Hank) Vazquez likes this

Joan

Joan Murt Llumà
R+D manager in Technologies For Advanced Manufacturing and Robotics
Top Contributor

@Paul, I can't agree you:

See Beckhoff, B&R, ABB, KUKA...

When a PLC is giving a life cycle of 10 years of spares and no new PLC's appear each year... Is easy to understand that Intel and AMD will make more powerful devices each... month? therefore it's easy to see that a quad core i7 processor will be faster than any XXX PLC processor out there.

Saying a PC is not as trustable as a PLC is like saying that ABB and KUKA products would not be trustable as they are using PC's to control their robots... the first one though is the first robot manufacturer in the world and the second one is the first manufacturer in Germany...

Unlike • Reply privately • Flag as inappropriate • 3 days ago

Enrique (Hank) Vazquez likes this

Paul

Paul Bennett IEng MIET
Systems Engineer at HIDECS Consultancy

@Joan;
I would not count the likes of ABB as a PC maker. So I think perhaps your definition of a PC is somewhat different to my definition.

Unlike • Reply privately • Flag as inappropriate • 3 days ago

👍 Enrique (Hank) Vazquez likes this

Joan

Joan Murt Llumà
R+D manager in Technologies For Advanced Manufacturing and Robotics
Top Contributor

Abb and kuka are robot manufacturers that use pure PCs to control their robots...
Unlike • Reply privately • Flag as inappropriate • 3 days ago

👍 Enrique (Hank) Vazquez likes this

Andrew

Andrew Pierro
Business Unit Director at Burkert Fluid Control Systems

Over time industrial PC boards will become more popular 15-20 years. But single loop and multi-loop process controllers are still strong in the market with new products coming out and they are no where near as powerful as a PLC. Why... pre-programmed very easy installation truly plug and play. The same is true with many PLC software, with function blocks, all starting to focus on making it easier and faster to set-up and program. So for 85% of industrial automation the PLC is not going anywhere fast and Rockwell and Siemens as well as a few others are not ready to switch their core focus.
Companies like ABB and KUKA it makes perfect sense for them to go PC, because the advance algorithms they have to write for the kinematic moves of their robots. Most all PLC's cannot realistically handle that type of computations. Plus all their machines programs are almost the same format.

The market will change when price, reliability and ease of use out ways the current options. With PLC sales still extremely strong and most machines are built with ten years + service life. They will be around for a long time.

Unlike • Reply privately • Flag as inappropriate • 3 days ago

Enrique (Hank) Vazquez likes this

Mark Froehlich
Energy Systems Architect at Efficient Automation and Controls - EAAC

YES - I'm using small footprint computers, 4x5 inch, 4 - threads, Embedded 7 to run Wonderware ArchestrA intelligent Objects using the PLC as distributed I/O and ..... more ..... contact me

Unlike • Reply privately • Flag as inappropriate • 3 days ago

👍 Enrique (Hank) Vazquez likes this

Robert Walter

Robert Walter Gomez-Reino, PhD
Senior Software & Systems Consultant specialized in large control systems

Hello @Paul, I personally agree with @Joan. At CERN, we started already designing more than a decade ago control systems for the LHC experiments based on PCs (call them servers if you prefer), and we are operating this systems since 2009. In the experiment for which I participated in the design, development and integration, we used about 100 (very powerful) servers for the control and automation of the gigantic experiment detector. Of course we had also many PLCs and other dedicated hardware but mostly doing safety last-resort related tasks. The automation and operation is otherwise done by PCs. These PCs host a control system taking very sophisticated decisions (statistical algorithms, expert system based, Finite State Machines control...) over about a hundred of thousand of controlled channels; the decisions are based on the acquisition of more than 3 millions monitored parameters. Rest assure we achieved extremely high reliability and availability. Among other things we made hardware (including some PCs) and software redundant sub-systems were we considered necessary.

At the end it is all a matter of cost, isn't it? You can probably make a PC system very reliable but to make it as reliable as a PLC you would end up spending much more than the cost of the PLC itself; so the question is how much reliability you need in each of the parts of the system and which of the options (PC or PLC based) can provide it at a lower cost. And when one talks about cost I mean also cost of maintaining the solution, cost of development, etc. spanning over time. When one talks about really big systems then things get much more interesting as you can create hybrid (PCs and PLCs) solutions where you have to optimize that cost while satisfying the safety, reliability, etc. requirements in a modular fashion.

Excuse me for such a long answer! I think it is an interesting topic as contrary to what many people think, PC-based automation (mostly mixed with PLCs as I said!) is out there already in industry too!

Unlike • Reply privately • Flag as inappropriate • 3 days ago

👍 Andrzej Czechowski, Enrique (Hank) Vazquez and 1 other like this

Enrique (Hank) Vazquez
☆☆☆☆☆ Your Plant Automation Concierge! ✔Control System Integrator ✔PLC Programming ✔PID Tuning ✔Consultant Engineer
Top Contributor

Great discussion, gentlemen. Thank you all for the insightful perspectives.

We can see similar lines being blurred between PLCs and SCADA platforms. The marriage of Big Iron and Big Data will produce many interesting control system iterations.

Delete • 3 days ago

Joan

Joan Murt Llumà
R+D manager in Technologies For Advanced Manufacturing and Robotics
Top Contributor

@Andrew, you are right when you say this won't be fast, even each day there are more people switching from the "old school" to the "new one", to change a market like the industrial won't be fast. :)

And yes, if you think on the PC as a standalone system you will have more difficulties than with a PLC to start doing things, but if you add a "soft-plc" system (like Beckhoff TWinCAT, B&R Automation Studio, KW proconos...) you'll be able to take the power of the PC using any programming language you'd like to: IL, FBD, ST... so in this case, you'll get the same easiness in implementation freedom of fieldbusses and a lot of power. Here you can use the advanced kinematics, calculations and plc code in the same controller. This solution would be something similar to an interpreter... like the JAVA virtual machine but completely dedicated to control IO, logic flow...

Unlike • Reply privately • Flag as inappropriate • 3 days ago

👍 Enrique (Hank) Vazquez likes this

Ganiev Dmytro
Automation and Instrumentation manager/engineer

Dear Joan, very interesting points from you. But I can't aggree. The rule for control system - make it simple and reliable, so you will perform changes to the existing system easier. Who will say that hot mix of C++ and .NET is simple, reliable an easy to change without unpredictable outcomes??? PC has it's niche and PLC has it's niche.

Unlike • Reply privately • Flag as inappropriate • 1 day ago

👍 Enrique (Hank) Vazquez likes this

Joan

Joan Murt Llumà
R+D manager in Technologies For Advanced Manufacturing and Robotics
Top Contributor

@Ganiev Dmytro, of course it is like you say: simpler equals better (always only if it is possible to get the job done); but re-reading the original question and trying to answer it: in certain advanced applications PC won't replace the PLC it already has done it.

Regarding your comment, of course everything depends on the background the user has. I truly prefer the power of one PC and the freedom of choice and to be able to create whatever I do need, whether CNC, PLC, HMI, remote connections.... given the kind of complex machines I have to deal with. When mixing CNC, PLC, artificial vision, robots and other technologies it is far advisable to get all the possibilities open and not being constrained by limitations.

Predictable outcomes are not as frequent as they should, I've programmed using different devices and brands (Siemens, Omron, Allen Bradley, Telemecanique...) and each of them come with different bugs and problems... The only rule that has worked for me always has been:
Test, test, test, when finished start again testing and create libraries that maximize the output and success.

Each tool for it's job, this is a good rule of thumb. But the right tool sometimes depends also on the field of expertise and background of the person who will have to use it.

I've never said that to use successfully a PC in an industrial environment you should use only .NET or C++, what I've said is that you could use a SoftPLC (see older posts) which would make the PC to work like a ultra-powerful PLC and CNC plus adding the opportunity to use any .NET or C++ programming environment to add rich features to your application (like a specialized CNC DIN code editor that would offer the operator assistants, syntax highlighting, real time guidance in programming...).

Without some help from a company like the ones I mentioned in older posts, it would be difficult to interact with inputs and outputs from a PC. Nowadays you can use the easy to find, fast and reliable Ethernet in any desired flavour (I use EtherCAT which is truly powerful) to connect your controller (i.e. PC) to any device (servoamplifiers, distributed IO, drives...).

Preferences, what is best for anyone... this thread was not about that. It was about the possibility of replacing a technology that has been out there for a lot of time for another one that offers different things. For me the choice is clear as I've been using it for long and never looked back, but of course this is my opinion.

Unlike • Reply privately • Flag as inappropriate • 1 day ago

👍 Enrique (Hank) Vazquez likes this

Ganiev Dmytro
Automation and Instrumentation manager/engineer

Ganiev

Software PLC have a lot of hardware compatibility restrictions and issues and real time OS required. Besides it's price is compareble with prise of real hardware PLC.
As I see new PLC systems are done on the PC-compatible hardware. But it still looks like PLC, smells like PLC...Though if you dig deeper its a masked PC. Maybe this is what you trying to say?

Unlike • Reply privately • Flag as inappropriate • 1 day ago

👍 Enrique (Hank) Vazquez likes this

Ganiev

Ganiev Dmytro
Automation and Instrumentation manager/engineer

I didn't work with robotics but for process controll automation it is very important to see program online and ability to dowload changes to PLC program without stopping the process. Is that possible with PC and compiled C++ programs? Process stop can be very expensive and sometime absolutely not possible!

Unlike • Reply privately • Flag as inappropriate • 1 day ago

👍 Enrique (Hank) Vazquez likes this

Ganiev Dmytro
Automation and Instrumentation manager/engineer

Ganiev

Sorry, after reading all post I lost the topic.
Do you agree with the prediction that PCs will replace PLCs in certain Advanced Automation applications?
Yes, I agree. Actually PLCs even didn't occupied some Advanced Automation applications.

Unlike • Reply privately • Flag as inappropriate • 1 day ago

👍 Enrique (Hank) Vazquez likes this

Jay

Jay Harbath
Contract Control System Engineer at Baxter International Inc.

PC based control still has a very long way to go to match the reliability of PLC based control. In over 35 years, I can count PLC processor hardware failures on one hand. I have NEVER had a blue screen of death equivalent software error on a PLC. And of course PCs are much more powerful than a PLC, but from what I have seen over my career is they are inherently unreliable. How many PC hardware and software problems have you had in your career? Are you willing to risk a billion dollar business on a house of cards built on a bed of Jello? Control system costs are trivial compared to downtime and product quality lawsuit costs.

Unlike • Reply privately • Flag as inappropriate • 1 day ago

👍 Enrique (Hank) Vazquez likes this

Andy

Andy Robinson
Principal at Phase 2 Automation

Wow! What an incredible discussion. I feel like jumping in not because I have something to add, but to relay some slightly alternate thoughts. At my previous job I lamented the idea that I had to pay north of $US15K for a Control Logix L73 with 32 or 64 MB of memory. If you consider that the "memory" is esentially just volatile RAM, how long do you have to go back to find RAM that expensive in a PC.. or going from 16 MB to 32 MB costing on the order of $US5K. I think the question to ask yourself is why we have PLC's that are 20 years old. Some will say it's wonderfully reliable and there is no need to change. Are you really saying that there is nothing in the new generation of controllers you would like to use so you are completely happy with what you have? Probably not. The "honest" answer is more along the lines of "It's still running and we don't have any complaints about the platform... or at least not enough to justify the re-engineering and hardware swapout cost". I have quoted a lot of PLC-5 to Control Logix jobs. Yes, AB has a nice upgrade path where you can reuse the IO or replace it with Logix cards in the same footprint. It is still very expensive and prone to error during the conversion process.

Here's another angle to look at.. security. If you have been following the Dale Peterson/Eric Byers debate about whether or not we can afford to rip and replace a bunch of controllers the answer I align with is that it is just too expensive to replace this PLC hardware with more secure models. Now, what about a model where all of the processing is done in a Server and the only thing outside of the server is IO. While some will still decry the security posture of a windows machine.. ok fine run a linux appliance if you want a smaller footprint...it is still worlds easier to securely communicate with a modern operating system than with a PLC of today's vintage. Yes you are still exposed on the IO front but at least you've cut the attack surface effectively in half.

For me having the gentleman from CERN tell us that they are using PC's (I know it's really servers....) in this fashion really just ended the debate. I seriously doubt anyone on this list can even sniff the order of magnitude, complexity, and criticality of what these guys are doing. If it's good enough for them then maybe the only places it's not good enough might be base regulatory and safety control in a Nuclear Generation facility or inside the cockpit of a Jet Fighter.

On a final note I think you can absolutely make a setup that would come very very close to matching the reliability of a PLC. I can take two machines on two seperate VSPhere servers and run them in Fault Tolerance mode. If one fails for any reason then the other one picks up instantly.. in milliseconds. For reference, in the not too distant past this was just Marathon EverRun under the hood. Or, run your environment on a Stratus box. If you virtualize I suspect you could fine at least 10-15 PLC "appliances" with 256 MB RAM footprints on there and the cost would be dramatically less than an equivalent fault tolerant setup.

I don't want to get distracted by cost though. I think what bringing cost into the equation does is give you the ability to look at superior function for equivalent money. What if we could run 20 separate PLC's and microsegment the control instead of putting everything on one or two pieces of hardware because you couldn't afford 20 controllers at $5USK a pop. I actually got this question from a previous customer at a very very large biologics facility. I think he was really just exploring the idea but it was a great thought exercise for everyone involved.

- Andy

Unlike • Reply privately • Flag as inappropriate • 1 day ago

Enrique (Hank) Vazquez likes this

Dr Cliff

Dr Cliff Jolliffe
Managing Director at Aerotech Ltd

Aerotech have been manufacturing a highly successful PC based CNC motion controller for over 10 years. It has so many advantages over its hardware based predecessor. We recently added a open plc (motionpac) to its platform and this has opened up opportunities in many new customer types. Would I say that there wasn't resistance to moving to a PC based PLC system? Absolutely. Secondly customers who have software running as a task on the motion controller where as the the PLC would serve them better, still have a preference to do it the old way. I guess the same applies to those considering changing from hard PLC to soft PLC.

Unlike • Reply privately • Flag as inappropriate • 1 day ago

👍 Enrique (Hank) Vazquez likes this

Duane

Duane Klein
Software Product Manager at T&H

To get back to the original question, "Do you agree with the prediction that PCs will replace PLCs in certain Advanced Automation applications?" Perhaps we need "Advanced Automation applications" defined.

It is clear to any one who works in the areas of business where PLC's have been used for control, that putting a PC for control on most small and medium OEM or skid based machines doesn't make sense. When you need a limited amount of IO, have very little space to work with, and need a 4 or 6 inch OIT, putting a PC in for control would not meet cost, space, simplicity, or reliability parameters.

As to the cost of those expensive PLC's, PC's don't hold a candle. When a customer tells me that it is time to migrate the Allen-Bradley PLC-5 that they installed and has been running since 1985, I think of all of the Microsoft operating systems that have come and gone in that same period of time, as well as the PC hardware changes that have taken place in that period of time. A PLC is called on to perform a specific function, to do it reliably, and cost effectively. If I divide the cost of a PLC system by the hours it just plain runs, and compare that to the cost I have seen associated with upgrading, and\or fixing a PC to do the same function, there is no comparison. How many PC's are any of using that we bought in 1985?

So back to the point of advanced automation applications. There are very few times in working with control systems over the past 30 years where I have needed the horsepower to run a particle accelerator and all of the associated calculations. To me, that isn't a PLC application. However, running a machine, managing a batch process, the balance of plant for a utility, or controlling a waste-water treatment plant, I would stick with a PLC because that is what they are designed to do.

Unlike • Reply privately • Flag as inappropriate • 1 day ago

👍 Enrique (Hank) Vazquez likes this

Grant

Grant Thomas
Building Automation / SCADA Developer & Operator

To answer the topic question: Yes, I do believe that PCs will replace PLCs in certain applications.

Firstly, the key word is certain, meaning not all. This allows a small company to use a PC based 'controller' and satisfy the question.

I come from working with PCs for almost two decades in my personal and professional life, and have spent many hours working on and with these PC systems.

Personally, I think that over time PCs will be doing a larger portion of the controlling of hardware processes. Remember that PC does not equal "Microsoft" or "Windows", and that vxworks is the base of quite a few of the hardware solutions that I have seen used in building automation. Tridium Niagara R2 is based on vxworks, as well as quite a few Notifire Fire alarm control panels that I've seen used in large buildings. In fact, the Notifire system is a generic mini-ITX board with some custom connectors on it. Truly, it could be replaced quite easily.
To address the stability concerns about PCs, you CANNOT compare the typical "windows" consumer experience to that of a PLC. To begin with, no PLC user installs games, office tools, scanners, cameras, etc etc etc into the PLC. THIS extra software causes the windows instability and crash happy environment that we all know and want to avoid. I have personally witnessed HVAC controllers that are PC based, running windows 95 / 98, and that have quite successfullly chugged on for over 15 years with no issues. Not one bluescreen, not one crash. This is because the system was only loaded with the needed software on top of the base system, and there were no users to do anything 'extra' to help things along. Please, let us not forget that many of the 'BSOD's or blue screen crashes are caused not by 'Windows', but by extra software. Device driver installations replacing system DLLs with altered versions to work with the driver wreaked havoc from 95 - Vista, I think, when they added the system level dll protection which gave the later windows systems so much more of a perceived stability boost. When the users overwrite the dlls that are system specific, the system restores them to the factory files, preventing many of the issues we've all seen too often.
To address the replacement timeframe on PCs that we use: We mostly do not replace PCs in the consumer space because they can't work. There are a few typical reasons, but the big ones I see are:

A) It's not fast enough any longer for the new suite of applications / games.
B) The system has been slow/sluggish for a long time with popups, and it's just faster and easier to buy a new shiny toy and move your pictures over to it.

The days are over when a PC needs a noisy power supply, an actively cooled heat sink, and a floppy disk. Now there are many SoC solutions that are able to be passively cooled with only the air internal to the installed space for the new PC, and that do not rely on the shell of the case for cooling, as other systems have done in the past.

Do not take any of this to mean that I think that the PC is superior to the PLC, a PLC is a fantastic tool for when it is needed. But so is a PC.

Beckhoff has a line of embedded windows controllers, and JCI also has a line of controllers that are windows based as well. Both of these platforms are quite robust and stable, and personally I've seen the JCI platform run for years with no issues, for precicely the same reasons that windows does not need to be unstable.

Unlike • Reply privately • Flag as inappropriate • 23 hours ago

Enrique (Hank) Vazquez likes this

Jim

Jim Bowman
CarolCon Enterprises, LLC | Principal Owner & Director; Systems Engineering

@Grant: Thanks for the PC stability and longevity comments. It is a favorite mythical issue, typically emerges by PLC sales folks that discover a potential customer is even considering a PC based control solution.

Also, I've seen seen more than one reference to Beckhoff Controllers in this discussion. Surely there are industrial PC vendors, including JCI that merit some reference too.

Thanks to all for your comments.

Jim

Unlike • Reply privately • Flag as inappropriate • 22 hours ago

Enrique (Hank) Vazquez likes this

Victor

Victor Wolowec
Profit-driven executive thriving in sales and marketing environments achieving sustainable, fast-track growth

I think the discussion is focused on "forms of processing" data. PLCs and PCs both process data but have different form factors built around them. I believe the ultimate answer is that the market will continue to evolve form factors to meet evolving manufacturing needs (whether it be PC or PLC). I think if you look at current PCs and PLCs, they are not what they used to be 5+ years ago and they will certainly change 5+ years into the future. The best question is, "What are the unmet needs currently in manufacturing?" and then identifying how to meet them with the best form factor currently existing or to be created in the future.

Unlike • Reply privately • Flag as inappropriate • 21 hours ago

👍 Enrique (Hank) Vazquez likes this

Harold

Harold Simpson
President at Repower Automation

I have a customer that uses PC based soft controllers (Siemens WinAC RTX) almost exclusively. Most of this is "low end" on the grand scale. I can tell from experience that trying to replace the PLC with a PC is simply asking for trouble. It is not so much the technology that is the issue as much as it is the "human factor" that is leading to trouble.

The first issue is that when something goes into a PC it tends towards the "bad old days" of the "black box". Now you need specialists simply to add buttons to an HMI (let alone changes to the control system).

Next you have the mentality of "its a PC so we can add this or that piece of software to the machine". A very recent example of this was where IT decided to add a new anti virus program and it basically shut down the controller.

Then there is the fact that most soft controllers run on top of Windows. Even with RTX extensions it is STILL Windows and anybody who has used a PC in the last 20 years KNOWS Windows will always have reliability issues.

The bottom line is that unless you are writing a complete custom controller as a black box, the PC has no place on the shop floor for anything other than supervision and data acquisition. Even then, I would still NEVER base such a solution on Windows.

Now what I feel would be an ideal way to build a complex system would be to use multiple dedicated PLC's for control, all "cross linked" via Ethernet in some sort of "mesh" networking/control strategy. This would be done in such a way that any one controller could take over control for a defective CPU. All of this would be "supervised" by a PC that also functions for HMI/Data acquisition and higher level dissemination.

This is all my personal opinion but my personal philosophy is that "distributed intelligence" is better and more reliable than "centralized intelligence".

Unlike • Reply privately • Flag as inappropriate • 13 hours ago

Enrique (Hank) Vazquez, Duane Klein like this

Jarosław

Jarosław Królikowski
Industrial Automation Engineer at Faurecia Wałbrzych S.A.

Going back to the root question: 'Do you agree with the prediction that PCs will replace PLCs in certain Advanced Automation applications?'
I am far from using the words 'will replace'. The PCs as we know them are quickly evolving and change. I am aware of advantages of both and above all I know how both of them work and what they are. For me the difference is more in definition then in reality. PLC is in fact a computer designed to easily connect some extension (I/O, communication etc.) cards (of course not always). While the ordinary PC is (what a surprise) a computer designed to easily connect some (e.g. graphic) cards. Both works thanx to some operating system, and here starts the difference. The system inside of PLC is (or should be) extremely reliable. But you might be mistaken thinking that MS Windows equals PC. You wouldn't believe how many PLCs uses CE, mobile, embedded or other specialized MS Windows version as an operating system.
There are of course hardware differences too but they concern rather standard equipment than technical limits.
Computer users use many different games ahhh sorry programs and need monitors to see the action, and keyboard to input data or control programs. It is of course far away from an operator that observes some panel and inputs process parameters for PLC program.
So it is not the matter of replacing the PLC in every application. But like the author asked 'in certain Advanced Automation applications'.
I think there are many situations where it would be great idea to build the system basing on PC utilizing its built-in flexibility. I am not sure if its operating system must be MS Windows. There are much more reliable ones like QNX.
And finally I think there always will be space and reasons to use just PLCs, connect them to SCADAs, use in DCS nodes, built in or replace with PCs. More and more PLCs have functionalities of PCs. They are equipped with ethernet, USB, VGA or DVI ports. Are they becoming PCs? No they rater won't ever be Personal but for sure they are Computers.
In my opinion PLCs will gain more and more advantages previously characteristic for PCs. I think we remember the times when computer system (if existed) on board of the plane (you flied) could control only the flight parameters and now the system also delivers some entertainment for passengers.
Regards
JK

Unlike • Reply privately • Flag as inappropriate • 1 hour ago

Eric Tam
Principal Systems Consultant at ITMation Ltd.

Eric

From my painful experience, a PC based system works great and is a cool, all-in-one solution, until the day that some hardware components in the box breaks down and you have to throw away the whole system because the manufacturer has stopped producing the product due to the lightning fast technology innovation cycle of PC. Then, rebuilding a replacement system is always painful, as you'll have to spend a whole day working out the hardware configuration and installation software for your new PC!

The beauty of a PLC system is that, if the processor breaks down, simply replace it with a new one and plug the old memory card to the new processor. Even if the original processor has become obsolete, converting the PLC program from an old processor to a new one is usually a much simpler process than rebuilding a PC system to interface with an existing control system.

My paradigm is, if the system needs to operates for more than two years, use a PLC system.

Unlike • Reply privately • Flag as inappropriate • 16 hours ago

👍 Duane Klein, Enrique (Hank) Vazquez like this

René Heijma
Product engineer Industrial Communication at Omron Europe BV

Could be that I'm an old fart in this business as I started to work with PLCs more then 25 years ago. At that point my mentor said: "Listen boy. probably PLCs will be around another 5 years but PCs will be the future!""
Currently I'm working for a company of which one of their main businesses is producing and selling PLCs. So there must be something good about PLCs.

So heard this before, but nothing changed.

Unlike • Reply privately • Flag as inappropriate • 7 hours ago

Enrique (Hank) Vazquez likes this

Robert
Walter

Robert Walter Gomez-Reino, PhD
Senior Software & Systems Consultant specialized in large control systems

OK. I am up for a round two :) I hope this is interesting for someone.

Gentlemen, it is not about getting rid of PLCs. What started this discussion was the question about computer-based controls taking over some of the tasks that PLCs traditionally did. Like I said before and others too, there is nothing to be predicted here, it has happened already. And this is not just about particle accelerators like someone pointed out, I personally know about many industrial systems where it also happened and others that they currently planning to do it. I just saw some weeks ago the control system of a national leader for oil product transportation. All of their terminals are automated and their control system has been already computer-based since more than a decade... I have no doubt (I didn't need to ask) that their systems has plenty of PLCs and other fail-safe hardware around; however, if their control system computer infrastructure would go down, no truck will load/unload oil. Guess what... it just doesn't happen.

Once again, it is all about requirements and the best architecture fitting them. People that have suffered when trying to use PCs for their control systems where probably in a situation where they shouldn't have had to. I can imagine hundreds of systems where my solution would be purely hardware based.

I have already seen a good set of very large control systems based in computers, many Linux, and also (YES!) a lot of them Windows-based too, and I am talking about Windows XP, Windows Server 2003, 2008... I personally designed the deployment and maintenance strategy for the control systems computers of a very large experiment. Those production servers have seen no blue-screens :). Please. These servers are installed by automated tools. From the operating system and its policies, the drivers, the SCADA software, to the very last script or hardware address connection. It all comes from a installation infrastructure we developed (SVN, configuration and equipment Databases, etc.). There is a clear overhead if someone wants to go to a computer-based reliable control system, this is not a standalone tower PC with a mouse and screen that we are talking about. These computers have never seen internet, and not even the engineers that developed the control sub-system that they are hosting have ever interacted with them. They are definitely stable systems. As for the hardware, well, you can buy today very fancy blade systems where the computers are sharing power supplies, network infrastructure and so on, you can install UPS systems, and in addition, you can always work on making your system computer and software redundant like we did.

The tendency today is in the direction of big system integrations. Standalone systems are the thing of the past. The bigger the systems gets, the more sophisticated cross-system decisions one can take, and the more profitable and effective is to take control intelligence to the upper layers (so farther from the hardware).

There is from my point of view today a huge gap from the people (please, no disrespect to anyone) that only sees automation as a PLC that opens and closes a gate and ont he other side the Web 2.0 gurus that want to control cities with their mobile phones. So here goes, after all @Enrique, a prediction: that gap will be filled by people doing controls and automation with computers.

Best regards!

Unlike • Reply privately • Flag as inappropriate • 6 hours ago

👍 Enrique (Hank) Vazquez likes this

Shaun Croman
Site Support Manager at United Biscuits
Top Contributor

Shaun

In reality they already have in some applications- Whether or not they should do? is possibly a better question.

Like • Reply privately • Flag as inappropriate • 5 hours ago

Harold

Harold Simpson
President at Repower Automation

Gentlemen,

Reading your fine comments and wisdom, has brought me to think of this from another point of view. Maybe the question should not be if the PC will replace the PLC, but more along the lines of will there be a shift away from centralized IO towards decentralized and modular IO.

I think that Mr. Królikowski brought a good point to the discussion. From a hardware standpoint there really is not much difference between a PLC and a PC. The real difference between the two is two-fold. The first is the method of (user) programming and the second is the method of accessing the IO.

Looking from the programming point of view, I think that a product such as Codesys is the future. Create a (more or less) common programming environment but a different compiler for different hardware. From all of the products I have seen and worked with, this is about the closest to "universal" that anybody has come so far. The problem though with its adoption is entrenched manufacturers of PLC's who do not want to see their fat margins disappear. Secondly, PLC "brands" are almost like religion, everybody has their favorite and for one reason or another does not want to shift to something "new". The same goes for "programming languages". Try to get an American programmer or maintenance person to use something other than Ladder! On the other hand try to get a German to use something other than Function Block Diagram (in case anybody is offended, I am simply generalizing to make a point). The point in this part of the discussion is it is more the "human factor" that should be questioned than the "hardware".

Next, looking at the IO, the only thing that really separates a traditional PLC from a PC (in the area of hardware) is backplane IO. In the past, the backplane was "nessasary" from the speed and reliability point of view. In this day and age though, there are a multitude of both fast and realiable "fieldbus" solutions. Traditionally these have either been CAN based or RS485 based. This "choice" has pretty much locked the machine builder into either the AB camp or the Siemens camp. Expanding on that, if a customer from Europe comes to the American machine builder, and asked for a "Siemens solution", the machine builder has had to redesign his machine control from DeviceNet to Profibus (assuming he even uses a fieldbus). Same goes for the Siemens to AB. Just this section alone I could probably write a whole book. For the sake of making an already long post, shorter, lets simply fast forward to the "future" and what my personal idea is for the machine of the future.

My idea of the future machine is devoid of the "religion of control manufacturers and languages". The CPU is simply a stand alone "box". That box can either be a "dedicated CPU" (AKA the PLC CPU) or a PC running some sort of controller software. Inside it is a program written with a very minimum of control manufacturer dependent code (AB-Siemens-Mitsubishi et al). The "skeleton" of the program would probably be written in Sequential Function Chart, where necessary, calling control manufacturer specific functions. The fieldbus is Ethernet based so as to have a common wiring. IO consists of a remote rack from any manufacturer that provides bus couplers for all of the Ethernet based fieldbus' (WAGO, Beckhoff, Phoenix, etc).

Under the outlined paradigm, a European customer now comes to the American manufacturer and asks for a "Siemens based solution" on his "AB based machine". All the machine manufacturer has to do is change the controller to one that supports Profinet in place of Ethernet/IP, and swap out the bus couplers for the same. Next the AB specific functions in the SFC program are swapped for Siemens specific but (for the sake of argument) 80% of the original program remains the same.

(continued in next post)

Like • Reply privately • Flag as inappropriate • 5 hours ago

Harold

Harold Simpson
President at Repower Automation

To summarize, let me just say that my opinion is, it is not so much a question of PC vs. PLC but more one of design methodology. The future (after we hopefully get past the manufacturer "religion" aspect of machine control design) will be one of "modularity" and "decentralization" as opposed to the current "monolithic" and "centralized" control paradigm.

Like • Reply privately • Flag as inappropriate • 5 hours ago

Grant

Grant Thomas
Building Automation / SCADA Developer & Operator

Harold,

Your description, and how I read it, of your vision of the future containing hardware agnostic code compiled onto the target machine sounds VERY familiar to the Linux world I come from. Personally, I use the Debian (I'm sure others offer the feature as well) distribution which allows for quick, easy, and mostly OOB method of cross compiling a project or set of code to a hardware architecture (ARM, x86, PPC, etc) either the same or different than the machine doing the compiling.

I would not be surprised if in the near future there are great tools that allow for using a user desired contemporary language which is then cross compiles into the 61131 or other field device native code. Yes, I'm aware that this stands the potential to add a lot of complexity until much of the bugs would be hammered out, but it would also provide more accessibility (i think) for programmers who are not used to the architecture to be able to successfully code and develop for the logical functioning of the system, instead of programming to the hardware. I kind of envision it sort-of-but-not-quite like HTML being a 'web natvie' code, which is generated by PHP, Ruby, asp.net, and many other html preprocessors who's output is what all of our browsers use.

So far as your I/O centralization point, I think that a centralized I/O environment will not go away, but that decentralized I/O will spread as far as it is feasible and that the impact of a particular I/O's parent going down does not have too high of an impact on the operation of the system it is a part of.

I do agree with you that the future appears to hold a situation that is in favor of modularity and decentralization among controllers and I/O.

As a small aside, I'm an American who programs PLCs as needed, and my go to is ST. Of course, I come from also developing with PHP and other syntactically similar languages, so ST was incredibly easy for me to pick up and run with. :)

Like • Reply privately • Flag as inappropriate • 5 hours ago

Harold

Harold Simpson
President at Repower Automation

Hi Grant,

Thanks for your comment! Actually, what you describe in your first couple of paragraphs is what Codesys does now. They create the runtime engine for the hardware in question be it ARM, x86, 8051 but have a common 61131 programming environment. The code is then compiled and loaded into the target and executed.

I think at this point though, the biggest thing holding back real "progress" in the world of controls is not so much the technology but what I like to call the "religion". This is especially evident in the whole "Allen-Bradley versus Siemens" debate. In the end it is really no different than "Ford versus Chevy". Everybody has their reasons for liking one over the other but when brought down to the "lowest common denominator", they both do the same thing just as well and the only real difference is what a person is used to (or which companies rep's buy them the best lunches :-)).

Lastly, concerning languages, I look at the available languages as tools in a toolbox. A person or company that only programs in ladder is like a mechanic that only uses a crescent wench. A good "mechanic" uses the right "tool" for the right job. ST is WONDERFUL for communications or large scale data manipulation. I use it primarily where a maintenance person will not need to know or understand how or why a block works and I am manipulating a large amount of data. On the other hand, I would use FBD or Ladder (equivalents in my opinion) for simple clear logic that a maintenance person may need to troubleshoot. Lastly SFC is wonderful to control the whole system and makes very clear where the program may be hung up and why. Even IL has a place when you want to get down into the nuts and bolts of what is happening in a very terse and logical manner.

This actually brings me to a pet peeve of mine. It never ceases to amaze me how many PLC programmers do not even comprehend Boolean logic. This are, in some cases, people with decades of experience who cannot understand how to code if you give them a logic equation such as X AND NOT Y = Z or how to mask and shift to form a 16 bit word from a byte and 8 bits (for example). On the other hand, there are the Computer Science majors who do not know to use a NC sensor for a high level sensor in a tank so as to guard against wire break.

Sorry if I have gotten a bit off topic here.

Unlike • Reply privately • Flag as inappropriate • 2 hours ago

Paul

Paul Bennett IEng MIET
Systems Engineer at HIDECS Consultancy
Top Contributor

Perhaps Enrique should have clarified the definition of PC and PLC when he started this thread. I have seen that some respondents definitions of a PC differ from mine. A PC is a personal computing device (the desk-top box, tablet and smart-phone can fit in here as well). When it comes to controlling plant the controller should have a much more focused purpose and be constrained to fulfil that purpose without the side-tracking of inappropriate applications. I would also expect that the controller is robust enough to exist in the environment in which it is working. There are some controllers that have full GUI HMI available to the user but that is just one processor in the collection (usually they have other processors that do the interfacing to the plant, quite often in the same box). Some PLC's have a very viable web-server interface that you can easily access from the Ethernet connection. They may be able to run Windows or Linux or other very capable GUI OS's and be able to lend themselves to running some of the office apps if you want but, if the device was networked, why would you want to?

My control world is in the higher integrity control arena. All processors I use are designed into the system to operate in harsh environments (vibration, temperature, humidity, magnetic fileds, intense transients, RF fields and, sometimes neutron rich zones, and have to operate over extensive lifetimes (25 years +) with little or no maintenance. I do not see consumer grade PC's being able to cope dependably in such an environment over such extensive periods. As there are plenty of applications in such environs I do not see the PC (compliant with my definition of one) as capable of gaining that strong a foothold there.

Unlike • Reply privately • Flag as inappropriate • 1 day ago

👍 Enrique (Hank) Vazquez likes this

Eric

Eric Tam
Principal Systems Consultant at ITMation Ltd.
Top Contributor

After reading all the threads and absorbing the wisdom of all walks of automation, I can see the trend that the hardware platform for industrial, embeded PC is getting closer and closer to that of PLC and the PLC's bar of performance is also being raised higer and higher, but probably would never challenge that of PC.

I would say that PC would be an excellent choice for complex tasks that require lightning fast execution time such as CNC machinaries, which a PLC would never make into such arena. However, these complex machinaries are often performed in air-conditioned rooms where space and power is not so much of a budget issue and the process is often batch based and manned. If the PC breaks down, it would probably just interrupt a batch and a technician is always there to replace the failed hardware.

However, if you're talking about mission critical processes where everything is cramped into a tiny control panel that is placed outdoor or inside a tunnel, unmanned, controlling an uninterruptable continues process, then you would probably feel more comfortable using a PLC, even in an event driven synchronized redundant configuration, which is unheard of in a PC world.

The beauty of a PLC system is that, its hardware and operating system is designed to do one thing (probably a few), which is to execute the automation program, whereas a PC system is designed to perform a million types of task that are not required in a down to earth automation system, hence a PLC has a lot less undiscovered bugs in the operating system and hardware to breakdown.

The bottom line is, if you need to do fast and complex tasks, then get a Nissan GTR (high performance without an unreal price tag). If you need to perform rugged tasks with a near bullet proof machine and lightning speed is not such of a concern, then get a Caterpillar tractor!

Unlike • Reply privately • Flag as inappropriate • 1 day ago

👍 Harold Simpson, Paul Bennett IEng MIET and 1 other like this

Paul

Paul Bennett IEng MIET
Systems Engineer at HIDECS Consultancy
Top Contributor

...but then if you need the performance and the ruggedness together, find the budget to build youself a Knight Industries 2000 (K.I.T.). :-)

Unlike • Reply privately • Flag as inappropriate • 1 day ago

👍 Shaun Croman, René Heijma like this

René Heijma
Product engineer Industrial Communication at Omron Europe BV

@Paul Bennet
Ah, those where the days.

Unlike • Reply privately • Flag as inappropriate • 1 day ago

Grant Thomas
Building Automation / SCADA Developer & Operator
Top Contributor

Grant

Eric, Paul:

I think I agree that the definitions of the subjects need a bit of work, PC may just be too generic. To me, PC refers to the hardware platform that it is built on. For example, x86, x64, ppc are the big players, and the ARM chipsets and architecture is rapidly gaining ground. When I see discussions talking about PCs, rarely is it that I consider it a requirement for a multiuser human interacting OS.
I think this is part of what sets my perspective on the situation. In the event that I were looking at setting up a PC platform for controls, my first instinct would be to use a trim RTOS, and as requirements demand slowly move up the chain to more typical user recognizable OSs such as bsd, linux, or even windows.
This then raises the question of dev time, specific application testing, troubleshooting, commissioning, and everything that goes along with it, and whether it's truly cost effective or even worth it to not use a PLC. But as has been said before, the right tool for the job is what needs to be used.

Harold,
This ties in to your statement of the ford-chevy style debates. IF we consider a PC a viable platform for taking the place of PLCs in at least a subset of the market, we will run into this issue on the OS side alone. You can see it in the more traditional consumer space with the 'windows', 'mac', or 'linux' shops, where they've picked their flavor and it's their way or the road.
Not that it's necessarily a bad thing, many large organizations have a slew of quite skilled technicians that are very good at bringing their standard hardware back to life from death (or close to it), and have stockpiles of same-model spare parts on the shelf that are as plug and play (for lack of a better term) friendly as any modular IO on a PLC.

Further, if you're considering a PC for a replacement of a PLC, then there are quick and easy methods of allowing for quick deployments. for PLC replacements with small storage memory footprints, a small CF or uSD card is a breeze to image and boot with almost 0 extra configuration, and for a few PC HMI/control systems that I've had input into, I have packaged the unit with a complete spare backup storage device that can be quickly installed to bring the machine back to a clean, fresh deployment status.
With the great strides that are being made with small physical storage devices, there are a lot of possibilities.

And while I admit that having worked for more than my adult life on PCs, computer networks, and the like, I do not think that my favor of the potential for the PC replacements of at least a portion of the PLC market is personal experience biased.
I think it's the untapped potential of the crazily extensible and extendable platform that it provides.

Speaking of going off on a tangent:
I wonder how much of the idle computing power of user's workstations and idle server time could be spent doing large data analysis and control system modelling for an organization, much the way that Folding@home and SETI allow spare computing time to be potentially put to good uses.
I do very strongly agree that there needs to be an isolation of controls and non-controls devices on networks, but maybe there is a place for the the others that we work with to help us help them, depending on what we're doing.

Unlike • Reply privately • Flag as inappropriate • 1 day ago

Jim

Jim Bowman
CarolCon Enterprises, LLC | Principal Owner & Director; Systems Engineering

@Paul: I concur that your PC would not be a good physical fit for most of the applications mentioned throughout this discussion; however, there are many commercial-off-the-shelf (COTS) industrial PCs (iPCs) that will and some are mentioned in the discussion.

@Harold: I really liked your comments regarding distributed control and I/O. During a recent assignment by an Oil and Gas Industry Midstream Developer, my team created a DCS consisting of nine (9) Controller Nodes distributed across about six (6) miles of "self-healing", Ethernet-over-Fiber Optic network (Controller LAN). Each Controller (Node), in this natural gas storage facility, is a Class 1; Division II rated iPC (Nematron P/N nPC300/ATOM N270 fan-less processor with SSD and Linux OS) contributes to all sequential, regulatory and safety control requirements of the DCS. Each Node interfaces with real-world I/O by way of some very intelligent Ethernet I/O subsystems (Opto 22 SNAP Class 1; Division II I/O, with EB1 Brains). Furthermore, each Node's I/O is distributed across its own I/O LAN ("self healing" Ethernet over Fiber Ring). The Controller Ring also hosts five (5) strategically located HMIs (Windows Workstation OS), a Plant Historian and a SCADA Data Server (both Windows Server OS). The run-time environments for all Controller Nodes' and HMIs' as well as the Historian and SCADA Data Server are configured by the user configurable open system (UCOS) Software of the FMC Technologies, Inc. and the system integration was completed by their own Automation and Control Group. I only mention this project because it illustrates yet another valid example where "industrial" PCs safely and efficiently perform some very complicated and critical tasks. Thanks again to all for your contributions, and the sage advice regarding safety and reliability.

Best Regards, Jim

Unlike • Reply privately • Flag as inappropriate • 1 day ago

Paul Jr Robitaille
Vice President R&D Maintenance at Granicor

Generic PC should have replaced PLC decades ago. PLC are overpriced pieces of equipment that fails anyway. I enjoy to see Beckoff (Twincat) engaging at full trottle into that direction.

Unlike · Reply privately · Flag as inappropriate · 18 hours ago